Structures

by Maryellen Gregoire

Consultant:
Adria F. Klein, Ph.D.
California State University, San Bernardino

capstone
classroom
Heinemann Raintree • Red Brick Learning
division of Capstone

People build structures to help them live.

Animals build things to help them live too.

People build apartments and live close together.

Cliff swallows build nests very close together.

People build stadiums.

Ants build hills.

People build bridges.

Spiders build web bridges between two plants.

People build tunnels.

Prairie dogs dig tunnels.

People build towers.

Termites build towers that can be as tall as a tree.

People build dams.

Beavers build dams with sticks and mud.

Can you build a structure?